# Cool Gadgets, Devices, Contraptions, and Tools

*Ingenious Innovations Unleashed!*

by

**Owen Jones**

## *Copyright*

Published by Megan Publishing Services
https://meganthemisconception.com

Hello and thank you for your interest in this book called *Cool Gadgets, Devices, Contraptions, and Tools - Ingenious Innovations Unleashed!*

Embark on a captivating journey into the realm of ingenious gadgets and cutting-edge devices with *Cool Gadgets, Devices, Contraptions, and Tools* As a gadget enthusiast, the allure of innovative technology is a constant source of fascination and excitement. This book serves as a guide to the latest and most intriguing gadgets, offering insights into their functionality, applications, and the impact they have on our daily lives.

From the convenience of smart home technology to the thrill of wearable fitness trackers and the awe-inspiring capabilities of drone technology, this book explores a wide array of gadgets that have revolutionised the way we live, work, and play. Whether you're a seasoned gadget aficionado or just dipping your toes into the world of innovation, there's something for everyone within these pages.

Discover the convenience of smart thermostats and lighting systems that adapt to your preferences, creating a comfortable and energy-efficient home environment. Dive into the world of wearable fitness trackers, which not only monitor your physical activity but also motivate and inspire you to achieve your health and fitness goals.

Explore the fascinating capabilities of drones, from hobbyist drones capturing breathtaking aerial footage to professional-grade drones revolutionising industries like agriculture and film-making. Dive into the world of cool car accessories and cell phone gadgets that enhance your driving experience and keep you connected on the go.

Join us on this adventure through *Cool Gadgets, Devices, Contraptions, and Tools,* where we delve into the world of gadgets, devices, contraptions, and tools that continue to shape the way we live in the modern world. Whether you're seeking practical solutions or simply intrigued by the latest technological marvels, this book is your guide to the ever-evolving landscape of innovation.

The information in this ebook on various types of gadgets, contraptions, gizmos, tools and adults' toys is organised into 21 chapters of about 500-600 words each.

I hope that it will interest those who like modern technology and gadgets.

Thanks again for purchasing this book,

Regards,

Owen Jones

## *Six Interesting Facts About Gadgets And Gizmos*

As a fellow gadget enthusiast, here are six fascinating facts about gadgets and gizmos:

1. The First Computer Mouse: The computer mouse, an essential tool for navigating digital interfaces, was invented by Douglas Engelbart in 1964. It was made of wood and had two metal wheels underneath. This early prototype paved the way for the modern mouse we use today.

2. The Origins of the Smartphone: While we now take smartphones for granted, the first recognisable smartphone, the IBM Simon Personal Communicator, was introduced in 1992. It had a touchscreen, calendar, email functionality, and even the ability to send and receive faxes.

3. The Evolution of USB: The Universal Serial Bus (USB) has become ubiquitous for connecting devices to computers. The concept was developed

by a consortium of companies in the mid-1990s. The first USB 1.0 specification was released in 1996, offering speeds of 1.5 Mbps. Today's USB 3.2 can transfer data at speeds up to 20 Gbps.

4. The World's Smallest Computer: In 2018, IBM unveiled the world's smallest computer. It measured just 1mm x 1mm and was designed to help track objects and monitor environmental changes. Despite its tiny size, it had the computing power equivalent to a 1990s-era PC.

5. The Rise of 3D Printing: 3D printing technology has rapidly advanced, allowing for the creation of intricate objects layer by layer. The first 3D printer was invented in 1983 by Chuck Hull, who patented the process known as stereolithography. Today, 3D printers can create everything from prosthetic limbs to architectural models.

6. The Smartwatch Revolution: Smartwatches, which combine the functionality of a traditional watch with smartphone-like features, have become immensely popular. The concept of a smartwatch was first introduced in the 1970s, but it wasn't until the 2010's that they gained mainstream success with devices like the Apple Watch and Samsung Galaxy Watch.

These facts highlight the rapid evolution of technology, from the humble beginnings of the computer mouse to the cutting-edge world of 3D printing and smartwatches. As gadget enthusiasts, we continue to witness the incredible innovation that shapes our modern world.

## *Table Of Contents*

## *1. Exploring Smart Home Technology*

In today's digital age, technology is revolutionising every aspect of our lives, and nowhere is this more evident than in our homes. Smart home technology, once a futuristic concept, is now a reality, offering a plethora of gadgets and devices designed to make our lives easier, safer, and more efficient. From smart thermostats that learn your preferences to security cameras that keep watch over your property, the possibilities are endless. Let's delve into the world of smart home technology and discover how these innovative devices are transforming the way we live.

Smart Thermostats: Efficient Comfort Control
Imagine a thermostat that not only controls the temperature of your home but also learns your habits and preferences to optimise energy usage. That's the magic of smart thermostats. Devices like the Nest Learning Thermostat and ecobee SmartThermostat use advanced algorithms to create personalised heating and cooling schedules

based on your daily routine. They can even adjust settings automatically when you're away, helping to save energy and reduce utility bills.

Intelligent Lighting Systems: Set the Mood with a Tap

Gone are the days of fumbling for light switches in the dark. Smart lighting systems, such as Philips Hue and LIFX, allow you to control your lights with a simple voice command or tap on your smartphone. Create custom lighting scenes for different occasions, from movie nights with cosy dim lighting to vibrant hues for a party atmosphere. Not only does this add convenience, but it can also enhance the ambiance of your home.

Security Cameras: Keep an Eye on Things

When it comes to home security, smart cameras offer peace of mind like never before. Brands like Ring and Arlo provide high-definition cameras with motion detection and two-way audio. Receive instant alerts on your phone when there's activity detected, and remotely view live footage of your property. Whether you're at work or on vacation, you can check in on your home and even speak to visitors through the camera's speaker.

Voice Assistants: Your Personal Home Concierge

The rise of voice assistants like Amazon Alexa and Google Assistant has brought the concept of a smart home to new heights. These virtual assistants can control all your smart devices with simple voice commands. From adjusting the thermostat to playing your favourite music and even ordering groceries, voice assistants make home automation effortless. Just say the word, and your wish is their command.

Energy Efficiency: Saving the Planet and Your Wallet

One of the significant benefits of smart home technology is its focus on energy efficiency. Smart thermostats optimise heating and cooling to reduce wasted energy. Smart lighting systems use LED bulbs that consume less power and can be programmed to turn off when not needed. By monitoring your energy usage in real-time, you can identify areas where you can cut back, ultimately saving money on utility bills while reducing your carbon footprint.

Convenience and Comfort: Tailoring Your Home to You

Above all, smart home technology offers unparalleled convenience and comfort. Imagine arriving home to a warm, well-lit house without lifting a finger. With smart devices, you can automate routines and schedules to fit your lifestyle. Whether it's waking up to a gentle sunrise simulation or setting the perfect temperature before you walk through the door, these gadgets adapt to your needs, making daily life more enjoyable.

Smart home technology is not just a luxury; it's a transformative way to live. From enhancing security and energy efficiency to adding convenience and comfort, these devices are reshaping the modern home. Embracing these innovations allows you to create a personalised, efficient, and safe living environment. The future is here, and it's smarter than ever.

## *2. The Power Of Wearable Fitness Trackers*

*Your Personal Health Companion*

As a gadget enthusiast, staying on top of the latest trends in wearable technology is a must. One of the most exciting and impactful developments in recent years is the rise of wearable fitness trackers. These innovative devices have transformed the way we approach fitness, health monitoring, and overall well-being. Let's delve into the realm of wearable fitness trackers and explore how these gadgets can revolutionise your approach to fitness and health.

The Evolution of Fitness Trackers: From Step Counters to Health Monitors
Fitness trackers have come a long way from simple step counters. Today's devices are sophisticated health monitoring tools that can track a wide range of metrics, from steps taken and calories burned to heart rate, sleep patterns, and even stress levels. Brands like Fitbit, Garmin, and Apple have paved

the way with their versatile and feature-rich devices, offering users a comprehensive view of their fitness progress and health status.

Smartwatches: More Than Just Timekeepers

Smartwatches have emerged as multifunctional companions that combine the features of a fitness tracker with the convenience of a wearable device. With built-in GPS, heart rate monitors, and activity tracking, smartwatches like the Apple Watch and Samsung Galaxy Watch offer a holistic approach to health monitoring. They also provide notifications for calls, messages, and calendar events, keeping you connected while on the go.

Health Monitoring Beyond Fitness: Insights into Your Well-Being

What sets wearable fitness trackers apart is their ability to monitor various health metrics beyond physical activity. Devices like the Fitbit Sense and Garmin Venu can track heart rate variability, oxygen saturation levels, and even detect irregular heart rhythms. This real-time data allows users to gain insights into their overall health and well-being, empowering them to make informed lifestyle choices.

Personalised Fitness Goals: Motivation at Your Fingertips

One of the most significant advantages of wearable fitness trackers is their ability to set personalised fitness goals and track progress over time. Whether you're aiming to increase your daily steps, improve your sleep quality, or reach a target heart rate during workouts, these devices provide real-time feedback and motivation. Some even offer guided workouts and coaching, making it easier to stay on track and achieve your fitness aspirations.

The Importance of Sleep Tracking: Enhancing Recovery and Performance

Sleep plays a crucial role in overall health and fitness, yet it's often overlooked. Wearable fitness trackers with sleep tracking capabilities, such as the Oura Ring and Withings Sleep Analyzer, monitor your sleep stages and provide insights into sleep quality. By understanding your sleep patterns, you can make adjustments to improve restorative sleep, leading to better recovery, mood, and performance.

Connectivity and Integration: Seamlessly Syncing with Your Life

Wearable fitness trackers are designed to seamlessly integrate into your daily routine. Many devices sync with smartphone apps to provide detailed analytics and trends over time. They can also connect to other health and fitness apps, such as MyFitnessPal and Strava, allowing you to consolidate all your health data in one place. This connectivity enhances the overall user experience and makes it easier to track progress and make informed decisions.

Wearable fitness trackers are more than just gadgets; they are powerful tools that empower individuals to take control of their health and fitness journey. From tracking daily activity and monitoring vital signs to providing personalised insights and motivation, these devices offer a holistic approach to well-being. As a gadget enthusiast, embracing wearable fitness technology opens up a world of possibilities for optimising fitness, improving health, and leading a more active and balanced lifestyle.

## 3. Buying Your Digital Camera

Digital cameras are here to stay, so whether you would like to take professional quality photos or merely a few vacation and family snaps a few times a year, you will have to become used to them.

The good news though is that the digital functions accessible are implemented in the same way on all digital cameras, so once you have learned how to use the contrast function on one camera, you will be able to operate it on all digital cameras.

This is not to say that these features are equally good on all cameras or that they will be accessed in the same way on all cameras though. A costly camera manufactured by a good manufacturer will almost certainly be better than a cheap point-and-click camera built into a mobile phone, but you would expect that anyway.

Most digital cameras have dozens and dozens of

features to control specialist aspects of lighting, most of which most happy snappers have no clue about and normally they do not want to know about them either. Many of these functions are available in picture manipulation software, so they are merely duplicated in the camera, where most people do not use them.

The first tip for buying a camera is not to choose it by its looks. Novices are often impressed by how the camera looks rather than what it can do. This is usually because they do not understand the features but they want their camera to be the size of a packet of cigarettes. So before you go to purchase a digital camera, take some time to acquaint yourself with the typical functions of a digital camera.

The first term to understand is megapixels. Digital images are made up of dots like a TV picture. The more dots the higher the resolution and the better the picture. The better the photograph, the dearer the camera. So, what type of quality pictures do you require and how much can you afford to pay?

Most digital cameras are crammed with features, but do you really want them all? If you intend using the photograph manipulation software that comes with most cameras, then you do not truly require the features built into the camera as well. If you do not need professional quality pictures to print off on paper, why pay for them? Simply buy a camera with only the features that you will use.

You can often purchase a less feature-rich camera in the sales, which are meant to clear out old stock before the latest models come out. The latest models will have more functions, so you can win all round by buying last year's model at a knock-down price.

Two functions that are worth having are a USB cable and connection and a memory expansion slot. The USB connection will allow you to easily upload your photos to your computer for manipulation and distribution and the external memory will permit you to take more photographs than the camera's RAM would usually allow.

## *4. An Introduction To Remote Controlled Helicopters*

It has frequently been said that every man needs a hobby and perhaps that applies to women as well, if they have the time, so I am going to suggest radio controlled choppers as a decent hobby. Getting interested in remote controlled helicopters is not the easiest manner of starting with remote controlled models, but it is probably the most challenging and perhaps the most fun.

Enthusiasts of RC cars and trucks would almost certainly disagree, but I am certain that there is a certain amount of cross-over too. The difficulty with starting a hobby with RC helicopters is that it is daunting, because the ultimate object is to construct your own flying machine with your own two hands and most individuals know very little about engines, nothing concerning aerodynamics in general and nothing concerning helicopters in specific.

So, how would you begin a hobby that has such a high-flying aim? The best place to start is almost certainly a mini RC helicopter. These mini RC helicopters are about eight inches long and weigh about three ounces, but they have three channel control and are very manoeuvrable.

They are also fairly cheap. Nowadays you should be able to buy one for about $30, which is around $100 than last year.

These entry level RC choppers come ready-to-fly, although you might have to affix the rotors, so you will not learn much concerning assembly here.

After you have flown your helicopter for a few days and you are beginning to understand it, get a subscription to an enthusiasts' magazine, so that you start to learn the terminology and different techniques.

You will probably have a few crashes and have to replace rotors or parts may wear out over time or be defective. They will need replacing. This is a good thing, because you will learn the fundamentals of assembly on a fundamental machine and the parts and tools are readily accessible.

After a time, you might be fed up of constantly recharging the battery, so you may decide to stop (so what? You had more than $30 of fun) or you may choose to move up a level and buy an RC helicopter kit. However, before you do that, go along to an enthusiasts' club and chat to other owners about the pros and cons of the different models.

At this point, you will be glad that you have been reading your magazine, because you will have read reviews and adverts on the latest kits and you should have some serious questions ready to enquire of the experts at the club. If there is no club near you, join a couple of the Internet forums on RC hobbies in general or RC choppers in specific.

Once you have your kit in your hands, do not be too eager to get to work on it. Read through the directions first and find all the parts. Some kits come with all the required tools others just supply the specialised items, so make certain that you have everything to hand before you begin and make certain that you recognise all the parts.

## 5. *Car Accessories And Gadgets*

Most drivers love gadgets and accessories and the chances are that, if you have a car, you do too. Drivers love accessories and gadgets from the modest dashboard compass or radio to the on board satellite navigation system and the sophisticated CD or even DVD player. They augment the enjoyment that you get from your driving experience and help turn your car into a copy of the comforts that you enjoy at home.

If you travel a great deal, driving can become dreary and so some accessories and gadgets are intended to alleviate that boredom although they should never distract the driver's attention from the primary job in hand, which is getting safely from A to B. I have listed a few of the most popular, non attention-diverting accessories and gadgets below so that you can see how many of them you already possess.

The number one, all-time favourite is the car radio,

although this was long ago upgraded to the car CD, DVD or even MP3. Few drivers would be without their car stereo. The car stereo can be a very practical tool as well, because you can set it so that, even if you are playing CD's, the radio will cut in if there is a significant road traffic announcement that will or could have an effect on you. This can be invaluable information, especially if you do not have the next item.

The next most popular modern device is satellite navigation, frequently called sat nav or GPS, which stands for 'global positioning service'. Sat nav can take the place of the radio announcement cut-ins and even go two steps further by suggesting an alternative route in order to avoid the obstruction and giving you instructions how to do it.

An essential piece of kit that every driver must have in the trunk is jump leads. Fan belts slacken, batteries short out, all kinds of items can happen, but if you want to get on the road again quickly, the best thing you can have if you have a flat battery is your own set of jump leads. It is no good depending on someone to stop and help you get your car started hoping that they will have jump leads. Get your own, they only cost a few dollars.

If you have kids that are a pain in the backside on

lengthy journeys, why not get them an in-car entertainment system? Strapped or cut into the back of the front seats, the screens can display video games or films - a bit like aeroplane seats, but more so.

If you like to drive fast but are fed up with speed cameras, why not fit a speed camera detector? They are not that expensive and can be fitted by yourself do-it-yourself. It could save you hundreds in fines and even your license.

In actuality, there are far too many car accessories and gadgets to be listed here and I am sure that you could cite a dozen that I have left out, but the point is merely that car accessories and gadgets can enhance your driving experience and that is what it is all about at the end of the day.

## 6. Cell Phones For Kids

Cell phones are not expensive now and, if you hunt around and do your research in the correct manner, so are local call charges. So should you include very young kids on your cell phone gift list? Well, it was not an option a decade or two ago, so this is a new dilemma - something else for parents to be concerned about.

Or is it? There was no option for cell phones for kids, because cell phones did not exist. Then until a decade ago they began to become cheaper. Nowadays, it appears hard to justify not giving your children cell phones. The main reason why parents like to give their children cell phones is for security.

Are there more risks for children now or does it only seem that way? Children have been walking to school for a hundred years, so why is that a difficulty now? Increased traffic is one reason given, yet inner cities were dangerous when horses and carriages were flying around corners too and there were fewer policemen and no traffic lights then either.

Or are we more frightened because the media loves to report bad news, so we are more conscious of these risks to children? Fifty years ago, my mother told me never to accept lifts from strangers. Why is that simple advice not enough any more?

Some people say that there are more predators than ever, because we have been poisoned by junk food, pollution and sex on TV. Whatever you think, there is no denying the fact that parents are very worried about their children whilst they are out of sight.

One pretty cheap manner of easing these concerns is to give your kid a cell phone. This is not a bad thing. It could even encourage your child to get into technology, because you can instruct your child how to personalise the phone with wallpaper, screen savers, ring tones, themes and profiles.

There are two routes to go with cell phones for kids. You could either get a pay-as-you-go phone so that the phone call budget is under your control or you could buy a cell phone family plan so that there are free calls between phones on the family network.

For safety reasons, it is best not to give your child a cell phone that is enabled to go on the Internet. It is also a safe idea to buy them one that cannot take photos. Most of the perceived threats to children come from the Net and the SMS (text messaging) of pornography.

If you buy a cell phone for your child, buy one that is just that - a cell phone, for making telephone calls in emergencies - and not one that is also a camera, video camera or an Internet console. A basic phone will give you the best of both worlds: safety for your youngster, without putting him or her at even more risk.

## 7. Airbrush Tanning

Since some people began thinking that a suntanned body makes them look trimmer, in all probability on the same principle that black clothing makes you look slimmer than white clothing, there has been a rush for instantaneous tans and there is no more instant a tan than to have one painted on. This is called airbrush tanning.

Airbrush tanning can appeal on different levels. It is quick, you can get any shade you like and there is no danger from UV rays natural or man-made. lots of people are frightened by the stories of skin cancer that they can get by sunbathing or using a tanning bed, so an instant golden bronze colour appears quite an attractive alternative.

The craze for airbrush tanning has spurred the tanning industry to manufacture an overwhelming range of airbrush-look-alike products. There are creams and lotions and even pills that the manufacturers claim will turn your skin an attractive golden bronze. Unfortunately, most of them turn your skin a rather ludicrous shade of orange.

Then there are the home airbrush tanning kits. Real airbrush tanning is similar to having your car resprayed. You can either have it done by a professional who has professional spraying apparatus or you could go out and purchase a dozen cans of your favourite colour car spray paint and do it yourself.

A car sprayed in the former way normally looks fantastic, but a car sprayed in the latter manner normally looks awful. Well, the same goes for professional salon airbrush tanning and home airbrush tanning. Even amongst professional spray painters some are better than others, so it is best to ask around before you let anyone airbrush you.

So, these are the greatest problems with airbrush tanning. You cannot do it yourself; you cannot rely on a friend to do it for you; it is doubtful whether you can even buy the right equipment to do it yourself and where do you find a specialist whom enough people have permitted to practice on them so that he or she is any good?

Because a decent airbrush tanner will need the correct equipment and plenty of experience. It is not a trade that can be studied from a book. Airbrush tanning is still a rather new phenomenon, so a good tanner might be difficult to find outside a big city, but you could try asking the owner of your local tanning salon to invite one in every fortnight for those who have booked in advance. A sort of guest appearance.

As with any paint job, the secret to a decent finish is preparation. If you have never been spray-painted before make sure that you prepare yourself in the correct manner. Ask your salon to supply you with a list of things that you can do in order to prepare your body properly.

Preparatory procedures may include shaving and exfoliating but will involve removing all make-up and body oils by showering well. This will give the paint a decent substrate to cling to. The better the preparation you do the better and longer-lasting the airbrush tan - ask any painter and decorator or car sprayer.

## 8. Cell Phone Tips

You have be conscious of the extraordinary rise in popularity of the mobile phone as they say in the United Kingdom or cell phone in the United States. Twenty years ago, having a mobile phone meant carrying a very expensive item the size of a house brick around with you and very few people owned one. Ten years ago, they were still fairly costly but a lot of individuals had one but now they are fairly cheap and even young children carry them about.

In fact, cell phones have become so popular in many countries, that landline telephone companies have started removing public telephones because no one uses them any more and they are too costly to maintain due to repeated vandalism.

It is very unusual to be seated in a public location like a train or a pub and not have someone near you talking into a mobile phone. In fact, numerous individuals get quite angry about people with mobile phones not becoming mobile when their phone rings.

Why do they not get up and go away to a quiet corner to take the call? No, most individuals merely sit there talking into their phone annoying everybody near them.

This then is my first cell phone tip: if you cell phone rings in a public location, get up and move away to answer the call. Everyone near you will be very grateful. We do not want to hear your private conversations and we do not want to feel that we have to keep our voices down because you are on the phone.

After learning this piece of cell phone etiquette, here are some tips for getting the most out of your cell phone.

This first thing to do when you acquire your mobile phone is to read the instruction booklet. If you always buy the same make, say, Nokia, you will be aware of the way that the phone works without reading the manual, yet if they have added new features you might miss them.

Manufacturers are constantly trying to remain one step ahead of their competition and they do this by adding features rather than reducing the price.

By actually reading the instruction booklet, you will become aware of these new functions.

Visit the manufacturer's website. There may be errors or omissions in the handbook or they might have upgraded the phone very recently. There will also be a help email address there for if you become stuck and an emergency phone number. The phone line will be either jammed or costly, so send your question by email instead.

Many cell phones allow you to assign different ring tones to different individuals. This is handy if used properly. For example, you could set one ring tone for your immediate family and the baby sitter, so that you know to pull over and take the call without even having to look at the screen.

Similarly, you could assign your boss a personal ring tone so that you can choose not to answer it if you are enjoying yourself.

## 9. *Baby Boomers And Hearing Aids*

The Baby Boomer Generation are the babies who were born in the first fifteen to twenty years after the Second World War, say 1945 to 1964. The countries of the world were pleased that the war was over and there was an uninhibited optimism in most regions around the globe, in spite of widespread devastation and the unparalleled loss of life.

This optimism for happiness, financial development and rising standards of living was passed from the parents to their children Baby Boomers. And the dream came true as well for millions of individuals. The Baby Boomers were well educated and had more wealth than any previous generation. Free Love and Flower Power came from the optimism of the Baby Boomers.

There were a number of other points that characterised the lives of Baby Boomers. Two of them were loud music both in the home and at live concerts and a greater inclination to consult a doctor because of their superior affluence.

The Baby Boomers are now beginning to get old. The oldest have even started retiring at 65 years of age in 2010. The next fifteen to twenty years will see hundreds of millions of pensioners join the queue for a pension around the world. Three million babies were born in the USA alone in 1946 and the boom became bigger and faster after that.

Years and years of warfare, loud music and riotous living will have had a harmful effect on our hearing. It must have had and I am certain that this is born out by millions of Baby Boomers looking for a hearing aid. Over the following twenty years this must mean that more and more individuals will be making appointments with doctors at the Ear, Nose and Throat Clinic.

Add to this the fact that hearing loss is also closely linked with ageing and the fact that we are widely forecast to live longer than our parents and you have a steadfast guarantee that the number of hearing aids required in the coming decades will rise sharply.

At the moment, hearing aids are still quite expensive. In fact, a good hearing aid costs more than a laptop computer and a decent digital hearing aid will cost more than two or three of the best laptops, However, if hundreds of millions of people getting ready to order a deaf aid, the prices will have to fall. And they will almost certainly fall a long way.

There are two conclusions you can draw from this. If you think that you will require a hearing aid but not right now, it would be a good idea to wait for a few years., if you can. The other is to put a bit of spare cash into the shares of a manufacturer of good, state of the art hearing aids.

Who knows, you could make enough money from your shares to buy a digital hearing aid 'free' or the company may even offer a big discount to its shareholders like quite a few of other progressive companies do already these days.

## *10. Cordless Phone Headset*

The cordless phone headset is a wholly hands-free unit and as a result permits the wearer the option to wander or operate away from the computer station.

This astonishing creation from Plantronics is the last word in wireless freedom for your administrative centre. The stylish Plantronics cordless phone headset boasts excellent sound quality and hands-free freedom of up to 100 from your desk - in a regular office, this means up to 50.

The Plantronics cordless phone headset is very light and comfortable to wear and can easily connect up to your present telephone system - it upgrades any traditional work station telephone into a fully wireless system.

It may be converted into any of three styles, so that you may make your mind up which is best one for you. The alternatives are: over the ear, over the head and (with an optional extra) behind the neck. The Plantronics wireless headset has up to nine hours of talk time and a rapid charge time as well.

You may even take telephone calls whilst you are away from your computer station, by way of their optional handset lifter - it makes taking a telephone call as trouble-free as picking up a headset.

## *11. Plantronics Cordless Phone Headset*

With Plantronics cordless phone headsets, you acquire the freedom of talking without a cord. Most of the Plantronics H-series headsets can be used as cellular headsets with the addition of a patch cable. Cordless phone headset units will typically have lower speaker volume because they don't have a headset amplifier.

With cordless phone headset units or standard phone headsets, the advantages of using hands free headsets and not having to kink your neck is a basic in today's busy office environment.

The H series includes the following Plantronics models:

H251: Plantronics H251 cordless phone headset continues to set the standard for durable, lightweight headsets for contact centre and office professionals. Expanding on the classic Supra style with new features, improved stability and updated styling, this H251 delivers a new level of all-day comfort and reliability.

HW251: model is the same as over but has a wideband speaker.

H91: H91 comfortable for all day wearing and is an excellent choice for the most discerning office professional

H261: the above have one ear speaker (monaural) the H261 two (binaural).

H101: the H91, but binaural.

H261-UNC: The Plantronics SupraPlus UNC (Ultra-Noise-Cancelling) wireless phone headset. The H261-UNC cordless phone headset is the gold standard in performance and comfort - even for the most phone-intensive uses.

A wider receive-side frequency response band, an intelligent flexible boom and chic design all bring greater audio accuracy and headset flexibility to the business professional.

HW261: Working with the latest in wideband VoIP technology, the Plantronics SupraPlus Wideband HW261headset delivers more advanced audio performance improving speech clarity, resulting in reduced errors, repeats and listener fatigue.

The noise-cancelling binaural HW261has a new extended boom to further reduce ambient noise. This list is not exhaustive, there are many makes and models. This selection is here only to give you an idea.

## *12. Different Types Of Auto Navigation Systems*

It is a strange fact that many buyers of new cars are ready to pay several times the true value of a satellite navigation system in order to have it installed by the maker. It is true that the car's manufacturer usually does an excellent job of installing the device, but then you are paying a premium for it.

If you purchased the sat nav unit separately and had it fitted by a third party, you would buy it for a third or a quarter of the cost. Still, it is part of the idea of buying a new vehicle to have all the latest apparatus built in to it. Satellite navigation, commonly called sat nav, is a real godsend, if you buy a system that is up-to-date and that is regularly updated.

It is not necessary to have the sat nav fitted in the car factory in order to have it fitted well. Many third party installers are quite capable of making a good job of it as well without having to have your radio/CD player removed.

Many auto navigation systems are fitted to the dashboard by means of rubber suction cups anyway. Buying a sat nav unit that does not have to have holes cut for it will also keep the price to a minimum without having to sacrifice quality or safety.

An important point to remember is that there are many kinds of GPS systems, each with rather specific applications. GPS for an ocean-going yacht does not need road maps, whereas GPS for a bicycle may not give enough advanced warning for the speed of a car.

Even if you buy a GPS sat nav system for a road vehicle, there are several varieties. The three fundamental kinds are: stand-alone, such as you see fitted at the car factory; hand-held and systems that are meant to be used with a laptop computer or similar set-up.

The stand-alone units are the most prevalent, because they have certain advantages: they are built for the job of getting you from A to B via C, D and E, if required; they hold a database of landmarks which will help you know that you are on the correct route; a voice will give you directions so that you do not have to keep referring to the screen and it will memorise and integrate previous routes.

Hand-held sat nav systems work, but require more thought and sometimes additional software to be supplied by the user. The screen is typically too small to be of much use and some only provide voice directions. Others only provide pictorial directions. However, they are better than nothing if you are walking or cycling in unfamiliar terrain.

Laptops and PDA's offer an excellent service, particularly if you already had the device for other purposes such as office work.

So, it is not just a question of getting hold of a cheap sat nav console and thinking that they are all the same, you have to see it running so that you can weigh up whether it is going to be of any benefit to you in your circumstances.

## 13. Ebook Readers

Do you read much? Do you spend a lot of money on books and magazines? Or do you go on holiday and have to heave half-a-dozen books with you to read at the poolside? If you do then it is worth considering buying an ebook reader.

A paperback may cost you $10, so if you take six books with you then that is $60, whereas an ebook reader will be between $120 $200, which includes access to millions of books and pieces of writing, lots of which are free.

The majority of ebook readers are a little bit larger than a paperback, but merely a third or the thickness. The reader can be preloaded with thousands of books so that you never have to be short of reading material. What is more, if you purchase certain readers, you will have Internet access to libraries of material to download everywhere in the world.

If you are an avid reader you will recover your outlay on the ebook reader within a year or so, because electronic forms of books are generally far cheaper than their hardcopy counterparts. Many ebooks cost less than $5 millions that have gone out of copyright are completely free.

If your ebook reader is enabled for the Internet and most of them have global Wi-Fi access, you can go online, browse the Net and read and send email as well. This is a colossal advantage as it means that you have a cut-down laptop computer at your disposal for little over $100.

That will save you having to visit an Internet cafe to keep in touch with your friends and family while you are on vacation. If you travel a lot, you will recover your outlay even more rapidly in this way. The ebook reader is also small enough to go in your pocket, briefcase or handbag..

Most of these ebook readers have a two-tone display, but there are more expensive devices that will also display in colour. For example, Kindle has a new device known as Kindle Fire for $199 is a colour ebook reader which will go online as any normal laptop will.

The big advantage is that as an Amazon customer,

you will have access to over 18 books, magazines, films and TV programs. This is the way that these ebook readers are going. They are in fact becoming complete entertainment devices.

Let's say you like a TV program and read a favourite magazine each week. If you go on holiday abroad, you will be able to watch the program and read the magazine before you get back home. These are quite fantastic benefits from what used to be humble ebook readers a few months ago,

Not just that, but say you start watching a program on the bus on the way home. When you arrive home, your electronic account manager will remember where you are in the film, book or TV program and you can pick it up from there on your large screen television.

## 14. Digital Camera Reviews

There are lots of reviews of digital cameras, in fact there are loads and loads of them both on and off line in the papers, in photographic magazines and on the TV. However, no matter how many you read, it does not make purchasing a digital camera any easier. Strange, isn't it?

Hopefully, the confusion comes from a healthy scepticism. We intuitively distrust a review if we imagine an ulterior motive and that motive is usually self-interest. Press releases from the manufacturer are the most obvious examples of this sort of ad.

However, there are others as well, like rehashed press releases made to look like personal points of view, supported by adverts selling the camera being reviewed. On line such rehashed press releases might be surrounded by Google Adsense ads or banner ads from retailers.

Home shopping reviews are next in line to not be trusted. The reviews that appear on Amazon might not have been tampered with my Amazon (and I think that they have not been), but who wrote them?

If I had a company producing digital cameras, I would make certain that home shopping sites like Amazon had at least two or three fantastic reviews of my cameras. Wouldn't you? You could compose three or four reviews in a working day and they would stay there on Amazon for the world to read for ever and Amazon ranks extremely well on Google - always.

Web sites can be dependable or not. They are the half-way house of reviews and you will need to use your critical judgment. The best indication is the advertising. Is the advertising surrounding the piece for the digital camera under review? If so, the review might not be fair-minded, if the adverts are for digital cameras in general, who knows.

Probably the most trustworthy digital camera reviews are written in magazines devoted to photography. These magazines build their reputation out of being independent and so have to uphold their credibility. These magazines protect their credibility like individuals protect their reputation, so cannot\page afford to be laughed at for pandering to a certain make.

Investing in a magazine or two with reviews on the digital cameras that you are contemplating buying is a small price to pay for getting independent advice which may save you money or stop you from buying a second-rate digital camera which might cause you years of anguish and aggravation.

Maybe the best reviews though are from friends and family members who have been using digital cameras for years. They will indubitably have been listening to other individuals talking about cameras for a time and reading articles every now and again as well. The only danger here is that people tend to talk up what they have bought because they do not like to look like having made a mistake.

## *15. Further Uses Of The Bug Zapper*

I don't know whether you have ever used a racquet-style bug zapper, but I think that they are awe-inspiring. I'm talking about the type that looks like a child's plastic, toy tennis racquet. They come in two basic kinds. I prefer the rechargeable bug zapper, because batteries end up costing more than the bug zapper itself, although you could always buy rechargeable batteries, but then they are dear too.

My wife and I like to spend time in the garden. We meet friends there, have a meal there and in general laze about outside, as do most people around here, when they are not working. Besides, it's far cooler outside than inside. A comfortable chair, some snacks, a chilled drink and a book or a companion and life does not get much better. In fact, it's idyllic.

That is until about six or seven o'clock when the first wave of mosquitoes have judged that the sun's rays have lost enough strength that they will not evaporate and they come out looking for blood. Some evenings are worse than others, of course. Usually, the mosquitoes are quite bearable, particularly seeing as I have discovered the bug zapper.

It's not that I like to slaughter things, but I find it hard to have sympathy for mosquitoes. Nevertheless, I do get a certain amount of enjoyment from seeing and hearing mosquitoes and other bugs literally explode with a flash and a zap as they come into contact with the charged and ground wires of the bug zapper. These bug zappers are capable of packing quite a charge, particularly if the batteries are new or the pack is fully charged.

The other day, I discovered a new use for my bug zapper. I'll tell you how it came about. I was in the garden, as usual, and my bug zapper was close at hand as the first wave of mosquitoes was due. I had my book in one hand and the bug zapper on my lap, when my wife asked me to go to the shop for her. No problem, therefore, I set off on the five minute walk.

I was half-way there when I realised that I had the bug zapper in my hand, but it was not worth taking it home and starting the trip again. Anyway, on my return trip, I had my small bag of provisions in one hand and the bug zapper in the other, when a local bully of a dog came running out of a garden straight for me.

This has happened often and, although he has never bitten me yet, it is rather menacing. He stood there glaring at me with teeth bared and his 'pack' of assorted neighbourhood friends came out to surround me and join in.

I don't actually know what the best course of action is in this position. I have tried standing my ground, but the intimidation just goes on and I have tried to continue walking, but he gets terrifyingly close by at times. This time, I all of a sudden lashed out with the bug zapper and just hit him on the nose.

Well, I'm not sure whether it hurt him, it did not appear to too much, but it gave him a very nasty shock in more ways than one, I can tell you! He leapt about four feet into the air as if he were on a pogo stick and then ran for all he was worth with all his friends following him. It was very gratifying after six months of persecution from this dog.

Anyway, I don't take my bug zapper all over the

place with me, but I will in future, if any further local dogs bother me. I know it works a treat. I have seen that one since, but he keeps far away from me and doesn't utter a sound. I believe I would take my bug zapper with me, if I were wandering in an unfamiliar part of town or the park nonetheless.

## 16. Tips For Buying A Car Alarm Online

Security is a huge problem these days. Everyone considers themselves at risk in some way or another and there are items on the market to alert one to these dangers. One such device is the car burglar alarm. Car burglar alarms have been about for quite a while, but nowadays they are very advanced. There are also tens, even hundreds of types of vehicle alarms.

Vehicle burglar alarms are a disincentive to most vehicle thieves or vehicle burglars, but they will never hinder the determined, expert thief. There are two types of car thieves. There are those who want to steal your vehicle and there are those who want to take its contents. Then there are opportunist car thieves and specialised vehicle thieves, some of whom steal to order.

In general, the more functions that a car alarm has, the more expensive it will be. Some vehicle alarms will have vibration sensors, some will have internal motion sensors and others will have immobilisers, which stop the engine from being started under certain conditions. Nowadays, most new vehicles come with a vehicle alarm built into the vehicle's electrical circuitry.

Some of these set-ups are extremely sophisticated, while others, usually less expensive models, are quite basic. The car alarms on a Bentley, for instance, use digital finger-printing, or biometrics, to open the door and turn on the ignition.

If your car did not come with a factory-fitted car alarm, you should add one as even quite sophisticated car alarm systems are relatively inexpensive now. However, most secondary auto alarms are fairly cheap because they still have to be fitted. If you cannot install a vehicle alarm yourself, the cost of installation is often more that the initial purchase price.

If you want to track down your own car alarm, whether you want to install it yourself or not, the cheapest place to locate one is on the Internet. However, because you will not have sight of the alarm before purchasing it, you should take additional care.

The first point is not to treat the Internet as an

opportunity to acquire the very cheapest alarm possible. It is a far better plan to use the Internet to purchase a model of vehicle alarm that you already know of at the lowest price.

There is a big difference. First find out which car alarm suits your vehicle and then go surfing for the cheapest retailer. When you have located a few places that sell the item you require quite cheaply, then you should look at the location of the retailer.

It can be a huge advantage to have the merchant near at hand, if there is a problem with the item. This can be even more of a catch with cheaper devices.

Next you should inspect the terms and conditions of the retailer's sales and returns policy, especially if the retailer is located in a country that is not your own. It is always safer to buy from your own country if it has good consumer protection laws.

## 17. Digital Camera Memory Cards

Digital memory cards are a camera's equivalent of a computer's floppy drive except they are stationary chips with no moving parts. Most digital cameras only have a small amount of internal memory, what is known as RAM on a computer, and this internal memory is really just for emergencies, because often it will only hold three to six photographs at the highest quality that the camera can produce. On the other hand, it might hold 50-100 quality photos.

While you are selecting a memory card for your digital camera keep in mind that not all makes of cards, frequently known as flash memory cards or flash cards, will fit into all cameras. If you do not remember which one you need, tell the shop assistant the manufacturer and model of your digital camera.

Once you have the right sort of memory card for your digital camera you can begin considering side.

However, there are one or two items that we should run through first. to help you realise why size is vital.

A digital picture is made up of dots of colour called pixels. The more pixels there are per square inch, the better the quality will be the photo. In other words, the photograph will have a higher resolution.

Another thing about these pixels is that some of them can record one of only a couple of thousand colours at a time and others can record one of millions, which makes for more accurate shades and tints - truer colours.

However, this higher capacity to record true colours comes at a price because every pixel has to have a larger amount of RAM allocated to it - one byte will allow 256; two bytes will permit 65,536; three bytes 16,777,215; four bytes 4,294,967,296.

These byte sizes are usually expressed in their bit sizes (eight bits is the equivalent of 8), so you have 8-, 16, 24 and 32. To place this into a perspective that may be more familiar to most individuals, Windows 7 in two versions 32-and 64-.

However, all these bits take up space, so the higher

the resolution you need for your photos and the truer you would like the colours to be, the larger the room you will need per picture. So, how good do you want your photos to be? Well, one question to ask of yourself is: what do I want to use the photographs for?

If you only would like to email them to your friends a lower resolution is better because it will send faster, but if you would like to print them out onto paper then a high resolution is better, particularly if you would like bigger prints. The larger the print, the higher the resolution the better.

So now you know how good you want your photos to be because you know what you are going to use them for, so the last question to answer is: how many photos do you expect to take? The answer to this usually depends on what you are doing.

If you are going on vacation, say a cruise putting in at five different ports, you may like to take five flash cards of 256 or larger and use one for each port. If you are going to one location, a card of 1 may be enough, but you can always take two or three.. If you are going to a wedding, you may like 3, 4 even 5 of memory, because you may like to print the photographs out

## 18. *Entertain Yourself With Cell Phone Ring Tones*

Everybody has a mobile phone nowadays. They all come out of their factories with their company's standard settings on them, so they all come out the same. All Nokias, all Sony-Ericssons are the same and all Samsungs are the same.

Millions and millions of them, all the same. However, they all, every single one of them have the potential of being customised in hundreds, if not thousands of combinations.

Most cell phones have a minimum of twelve different ring tones; twelve different screen savers, twelve different wallpapers, twelve different themes and twelve different colour schemes and some have twenty or more of every.

It is also possible to download additional variations and sometimes even to make your own. This means that there are normally thousands of ways of personalising your cell phone.

Although most people do not go to the bother of switching the wallpaper, screen saver or theme, most individuals do carry out trials with the ring tone. Older individuals may only use one of the standard ring tones that came with the phone, but younger individuals are more likely to either buy a ring tone or download one of the thousands of free ones available on line.

One difficulty for ring tone enthusiasts was (or even still is for some) that there is no real standardisation of the music format between the cell phone manufacturers. This means that a ring tone in Samsung format will not work on a Nokia et cetera. This led to the creation of ring tone converters, although some phones cannot make use of them.

Another way of overcoming this is to upload your ring tone to an on line converter, convert it and either SMS it back to your cell phone or download it to a computer and transfer it to your phone by wire or by Bluetooth. It is all too much trouble for most older people, although it is fairly easy when you have done it once or twice.

Other cell phones have a composer built in. These

composers are fairly basic in some phones, yet others permit polyphonic composition. However, it can get fairly tiresome to write a decent tune, so this is one for the serious customiser with time on his or her hands only.

One function that practically all cell phones have, and which most individuals do not use to the full is assigning different ring tones to different people or categories of individuals. For instance, you could have one ring tone for your parents or spouse and one for your home landline. Then if you hear those ring tones, you know that you should answer it.

You could have one for your friends, so that even without looking at the screen, you know that you could have an amusing conversation to brighten your day and you could have one for your boss so that you know to ignore it if it rings outside working hours.

It is entertaining to personalise your cell phone and it can be achieved fairly quickly once you get going and it allows you to while away half an hour next time you are waiting for a train.

## 19. Do You Need A Weather Radio?

There are radios that are especially for reporting the weather. Not everybody needs one of these dedicated devices, although we are all fascinated by the weather. However, the amount of information given out by most radio stations is sufficient for the majority of us. So what type of people would profit from a so-called weather radio?

Weather radios are most appropriate for people living in areas where extremes of weather can and do take place on quite an ordinary basis. If the area where you live is subject to hurricanes, tornadoes and flash floods or even severe storms, you are a likely candidate for a dedicated weather radio. Particularly if you have to travel away from home while an extreme weather event might take place.

All radio stations give weather news and weather warnings, but not all radio stations will suspend a programme to give 'stop press' updates on impending severe weather fronts. It is the same with television stations, not all of them will interrupt the highlight film of the evening to report on an approaching storm. Some of the smaller stations are not even subscribed to these types of weather reporting services.

However, it is not only people who live in regions of probable extreme weather who might benefit from these weather radios. People who carry out specialist activities and certain jobs need more specialised weather reports as well. For example, deep sea fishermen, sailors, farmers, mountaineers, hikers and backwoodsmen need to know if harsh weather is on the way.

A lot of weather radios are not only capable of broadcasting news about the weather. Many of them have a built-in AM/FM radio too and some will even act as alarm clocks. Some are mains only, whilst others are battery powered, wind-up or solar powered.

Some are bulky, but most are designed to be carried easily in a regular backpack and might have earphones too so that you can listen to a broadcast during a howling gale.

If you are just sitting at home, you might feel safe enough with the local TV or radio station on, but if you have to venture outside whilst there is a threat of severe weather, a weather radio is very comforting.

There are plenty of types and styles of weather radio to suit all needs, but a battery or wind up radio are the most reliable if you are away from a mains power source such as at sea or in the forest.

You will be able to find weather radios in a good number adventure or camping shops and in many chandlers. It is also straightforward to find these dedicated radios on line especially on eBay or Amazon.

Weather radios are not expensive to buy, but some models can eat up batteries so always take a couple of extra sets of batteries if you are going off the trodden track.

## 20. Top Five Auto Accessories

There are thousands of auto accessories on the market! just take a look on Amazon - it's shocking! However, what is equally shocking, depressing even, is that most of them are useless rubbish. Some are even dangerous.

Take the dip-sauce tray that clips on to your air vents, so that you can eat your junk food more easily while driving! Or the external camera monitor that is touch-screen, so that it is easily adjusted while you are motoring...

No, no, no! While you are driving, you should have both eyes on the road, not looking for dip-sauce trays, or adjusting touch-screens. You are supposed to stop your vehicle to carry out those tasks!

Then, there are the 'dodgy' gadgets...those auto accessories that make it easier for you to get as close as possible to flout the law. I have two in mind: 1] the breathalyser that 'is as accurate as those used by the police', and 2] the 360 degree laxer detector, that warns you when you are in danger of being caught for speeding. One, you should not be drinking any alcohol if you are driving, and two, you you are not the one in danger (of getting caught) if you are speeding, all those around you are. Don't do it!

I am going to assume that you have the ultimate auto accessory - a First-Aid Kit.
So, my choice of top auto accessories is as follows:

1] DASH CAM: I know that a lot of new cars come with a dash camera, but that only goes to underline how useful the motor industry thinks they are. For those who don't know, a Dash Cam will record the view out of the front (and, often, the rear) windscreen. This provides irrefutable evidence in the case of an accident. The best models can stream the footage to the internet, and record when the vehicle is being stolen or broken into.

2] JUMP STARTER: If your car is getting on, or if it has charging problems, a car jump starter is very useful. They are relatively small, and will also charge most electronic devices. One I saw had a powerful built-in light - handy if you need to use the device in a dark lane or garage. Some have enough charge for twenty starts.

3] DIGITAL ASSISTANT: If you are a fan of Amazon's Alexa, you will also, no doubt, get on with Chris, the auto digital assistant. Chris works like a human co-driver, in that it can give you directions, read messages and email, accept phone calls, and control the musical entertainment. it works off- as well as on-line.

4] MULTI-PURPOSE MOUNT: the best of these will hold any phone or tablet, and can be swivelled and locked into position. Most people fit them to the dashboard, which is fine, so long as they don't obstruct your vision. It helps turn your device into a usable GPS navigation system or an entertainment centre.

5] AIR FRESHENER: these things have become quite sophisticated. The cardboard pine tree hanging from the rear-view mirror is old-hat now. Many clip onto the car's ventilators, and use replaceable liquid canisters. Most firms have a variety of aromas - a bit like vaping, but the fumes are colourless. They are not only handy if you travel with pets, a windy partner, or eat food in your car, but they can help you stay focussed if you are travelling long distance.

There are also runners-up, which didn't quite make it onto my list like the high technology RFID tag you can add to your key ring. If you lose your keys, you can phone it to ask where they are; the car repair kit; the car bin; or the hooks you can fit over the back of each front seat's headrest to hang your shopping bags on.

## *21. Unveiling The Wonders Of Drone Technology*

As a self-professed gadget enthusiast, the allure of cutting-edge technology is always a thrilling adventure. In the realm of innovative gadgets, few hold as much fascination and potential as drones, or unmanned aerial vehicles (UAVs). These remarkable flying machines have revolutionised industries, hobbies, and even everyday life. Let's embark on a journey into the world of drone technology, exploring their diverse types, capabilities, and the impact they have on modern society.

Understanding the Types of Drones: From Miniature Marvels to Industrial Giants
Drones come in various shapes and sizes, each designed for specific purposes and applications. At the consumer level, hobbyist drones are compact, lightweight devices equipped with high-resolution cameras for aerial photography and videography. Brands like DJI and Parrot offer popular models

such as the DJI Mavic Air and Parrot Anafi, perfect for capturing stunning aerial footage and exploring the skies.

Professional Drones: Tools of the Trade for Industries

In industries such as agriculture, construction, and film-making, professional-grade drones have become indispensable tools. These drones are equipped with advanced features like thermal imaging cameras, LiDAR sensors, and GPS navigation systems. Agricultural drones, such as the DJI Agras series, help farmers monitor crop health, assess irrigation needs, and even spray fertilisers and pesticides with precision. In the construction industry, drones like the DJI Matrice series provide aerial surveying and mapping capabilities, reducing time and costs while improving accuracy.

Aerial Mapping and Surveying: Precision at Your Fingertips

One of the most significant advantages of drone technology is its ability to conduct aerial mapping and surveying with unparalleled precision. Drones equipped with high-resolution cameras and GPS technology can capture detailed images of landscapes, buildings, and infrastructure. This data is then used to create 3D models, topographic maps, and digital elevation models (DEMs). Surveyors and cartographers rely on drones for tasks such as land surveying, urban planning, and environmental monitoring.

Search and Rescue Operations: Saving Lives from Above

In emergency situations, drones serve as invaluable tools for search and rescue operations. Equipped with thermal imaging cameras and night vision capabilities, drones can locate missing persons in remote or hazardous environments. Organisations like the Red Cross and emergency response teams worldwide utilise drones to quickly assess disaster areas, deliver supplies to inaccessible areas, and provide real-time situational awareness to first responders.

Environmental Conservation: Drones as Guardians of Nature

The environmental benefits of drone technology are vast, particularly in the realm of conservation and wildlife protection. Conservationists use drones to monitor endangered species, track animal migration patterns, and identify illegal poaching activities. Drones equipped with sensors can collect data on air and water quality, detect deforestation, and assess the health of ecosystems. These insights are vital for informed conservation efforts and sustainable environmental management.

Recreational Drones: Elevating Hobbies to New Heights

Beyond their professional and industrial applications, drones have also captured the hearts of hobbyists and enthusiasts worldwide. Recreational drones offer an accessible entry point into the world of aerial exploration and photography. From racing drones that zip through obstacle courses to FPV (First Person View) drones that provide immersive flying experiences, hobbyists have a wide range of options to choose from. Additionally, drone racing leagues and aerial photography contests have emerged as popular pastimes, showcasing the creativity and skill of drone pilots.

Drone technology represents a thrilling frontier of innovation with limitless possibilities. Whether used for aerial photography, industrial inspections, search and rescue missions, or environmental conservation, drones continue to redefine the way we interact with our world. As a gadget enthusiast, embracing the world of drones opens up a realm of exploration, creativity, and technological advancement. Whether soaring through the skies or capturing breathtaking aerial footage, drones offer a glimpse into the future of aerial innovation.

## *Contact Details*

Facebook: AngunJones
Twitter: @owen_author
Blog: Megan Publishing Services

This book is part of the 'How to...' series of 150 manuals by Owen Jones.
The whole series can be found in many languages on Megan Publishing Services at:
https://meganthemisconception.com

www.ingramcontent.com/pod-product-compliance
Ingram Content Group UK Ltd.
Pitfield, Milton Keynes, MK11 3LW, UK
UKHW021647190726
13853UKWH00001B/108

9 788835 463887